Nikolay Seriukov

Combining Artificial Intelligence with TRIZ

Nikolay Seriukov

Combining Artificial Intelligence with TRIZ

TRIZ, ARIZ, AI and neural networks, versions of their combination in modern smart manufacturing

ScienciaScripts

Imprint
Any brand names and product names mentioned in this book are subject to trademark, brand or patent protection and are trademarks or registered trademarks of their respective holders. The use of brand names, product names, common names, trade names, product descriptions etc. even without a particular marking in this work is in no way to be construed to mean that such names may be regarded as unrestricted in respect of trademark and brand protection legislation and could thus be used by anyone.

Cover image: www.ingimage.com

This book is a translation from the original published under ISBN 978-620-8-11824-2.

Publisher:
Sciencia Scripts
is a trademark of
Dodo Books Indian Ocean Ltd. and OmniScriptum S.R.L publishing group

120 High Road, East Finchley, London, N2 9ED, United Kingdom
Str. Armeneasca 28/1, office 1, Chisinau MD-2012, Republic of Moldova, Europe
Printed at: see last page
ISBN: 978-620-3-29824-6

Contents

Keywords

Brainstorming, Innovation design, Quantum computer, Integral technical systems, Laws of development of technical systems, Vertical and horizontal integration, Function of analysis, Dialectical approach, Formulation of an inventive problem, Formulation of principles for achieving an ideal end result.

Annotation

The methods of modern innovative design are full-scale vertical and horizontal integration and the principles and methods of achieving the ideal final result should most likely turn into the principle of complex selection of autonomous functional components of the object for horizontal and vertical layout integration.

Since pure invention does not exist in the modern innovation process, at least one of the elements of integration should be the principle of commercial feasibility for horizontal integration and the principle of implementation universality in various technological categories, not always at first glance logically related to each other - for vertical integration and for linking the programme features of the incoming subsystems to the general concept of management and control.

Modern machines and devices, especially supersystems containing elements of artificial intelligence and artificial neural networks in the subsystems of control and monitoring have other technical characteristics, which form in a complex for the whole supersystem features - smart supersystem with all links between subsystems, defining - smart links and smart software mechanisms of interaction between subsystems.

Laws and practice of integrated development of integrated technical systems and supersystems in conditions of application of quantum computers and their simplified modifications in combination with built-in elements of artificial intelligence and artificial neural networks.

Due to the key factor that determines the efficiency of innovative undertakings and the correctness of the choice of patent and licensing strategy for the development of innovative projects with a high level of world novelty of technical and software solutions used, investors and managers of new (one can say - pioneer) projects face the lack of the necessary level of speed and depth of instant analytical processing of information when modelling processes on modern computers.

Information about the positive results of tests of the quantum computer created at the GUGL Corporation gave reasonable hope for progress in this direction.

CHAPTER 1

By definition and available basic information, -

A quantum computer is a computing device that uses the phenomena of quantum mechanics to transfer and process data

.(A quantum computer (unlike a conventional computer) does not operate with bits (capable of taking the value of either 0 or 1), but with Q bits, which have the values of both 0 and 1 at the same time.

Theoretically, this allows all possible states of systems at all levels (both supersystems and subsystems) to be processed simultaneously, achieving significant superiority over conventional computers.

In addition, the lack of such analytical tools currently significantly increases the cost part of the budget of innovation projects, as it requires significant expenditures on computer modelling and high-speed re-selection of options with analytical characteristic assessment of their acceptability and all aspects of efficiency.

Besides, the laws of technical systems development formulated in TRIZ (theory of inventive problem solving) cannot reflect the whole variety of tasks, functions and features of a modern multifunctional object, and taking into account all the new and emerging factors characterising an innovative object, it is necessary to redefine these laws, linking them with the laws of development of commercial structures and commercialisation of innovative ideas.

The formulations should be based on the obtained data and test results of quantum computers, taking into account the nature of the basic contradictions identified in the innovative object.

The dialectical approach (analysis of contradictions) embedded in the main tool of problem solving, which was ARIZ (algorithm for inventive problem solving), was distorted by the introduction of new concepts (technical and physical contradiction) for modelling of which in real time mode the speed and power of existing computers were insufficient.

These new concepts somewhat distorted the essence of the dialectical contradiction formulated in dialectical logic, which led to difficulties in identifying the contradiction when trying to solve real inventive innovation problems with the help of ARIZ, due to the lack of the necessary resource of speed and depth and scope of modelling processes and apparatuses.

We should focus on this separately, but there is an extremely important question of principle - what can be considered a real innovative inventive task?

From the point of view of an investor or project manager, analytical assessment of the project development requires a clear orientation of the situation and analytical modelling according to the following scheme: how a correct or incorrect formulation of the inventive problem may affect the commercialisation of the invention that has arisen?

Correct or erroneous formulation of project tasks and goals, in the absence of step-by-step modelling, can lead to misunderstanding of the classic question, - Is it possible to reliably protect the resulting technical solution from unauthorised copying?

The search for answers to all these and many other questions is now becoming a major part of the dialectic of creating a development strategy for innovative commercial projects and reliable patenting and licensing of inventions created within the framework of project implementation at all stages and phases of development.

During the period of its active development (the 1980s), these shortcomings and errors were successfully compensated by the enthusiasm of TRIZ adherents. Nevertheless, the existing flaws of TRIZ and the departure from TRIZ as a result of the production crisis of its main developers, who were able to see these flaws, led to a stagnation in the development of the theory. In our opinion, this is the main reason why nothing new worthy of serious attention has appeared in TRIZ in the last decade.

1. Crushing principle
*   to divide the object into independent parts;
*   collapsible
*   increase the degree of crushing of the object

The criterion of independence of parts into which it is recommended to divide an object in modern machines and apparatuses is practically impossible to fulfil.

Modern machines and devices, especially supersystems containing elements of artificial intelligence and artificial neural networks in the subsystems of control and monitoring have other technical characteristics, which form in a complex for the whole supersystem features - smart supersystem with all links between subsystems, defining - smart links and smart software mechanisms of interaction between subsystems.

If we take into account the fact that modern innovation objects are most often an integrative combination of apparatus, system, programme and method, it is clear that all parts or components of the object are to some extent tied to these elements.

If we follow this logic, it turns out that if it is necessary to achieve complete autonomy and independence of the object parts, it is necessary to endow each part with the correspondence to the specified ones - apparatus, system, programme and method, which, taking into account the requirement of the patent law on the indivisibility of the object of the invention, and knowing the design principle about the meaninglessness of repeating all the constructive, software, technological and algorithmic features, identified in the process of dividing the object into parts, in each of the parts, makes this technique not so much as a complete and independent method.

Methods of modern innovative design , is a full-scale vertical and horizontal integration and the principle of functional fragmentation is likely to turn into a principle of complex selection of autonomous functional components of the object for horizontal and vertical layout integration.

Since pure invention does not exist in the modern innovation process, at least one of the elements of integration should be the principle of commercial feasibility for horizontal integration and the principle of implementation universality in different technological categories, not always logically connected at first sight, for vertical integration and for linking the programme features of the incoming subsystems to the general concept of management and control.

The theory and algorithm for solving inventive problems were created at a time and in a situation when, at least in the Soviet economic system, there was no production of universal standardised structural and technological components.

This defines a fundamental difference between modern machine design and engineering and the design methods and criteria that existed at the time of the creation of the Theory and Algorithm of Inventive Problem Solving.

The recent emergence of all sorts of smart technologies, smart products, smart industries and their variants has created certain characteristics that are generally known and represent the main characteristics of innovative objects even before these objects become the subject of research and development.

Also, in today's conditions, taking into account the creation of independent modules and devices for non-contact control and non-contact speed measurements based on the principles of electromagnetic resonance spectroscopy, to divide the object into independent parts, in addition it is necessary to take into account the width and depth of coverage of adjacent elements of the supersystem requiring real-time control by the spectroscopy technique.

Analysing the brain response to intensive and active participation in system brainstorming aimed at intensifying the process of developing an innovative complex and modular project in the fields of smart technologies with elements of artificial intelligence and artificial neural networks.

Specialists constantly analyse brain reactions to brainstorming participation. Depending on the nature and specifics of brainstorming, brain areas are identified that are activated depending on the complexity and intensity of the psychological stress of the brainstorming process when experiencing the specifics of the technical characteristics, the required qualifications of the brainstorming participants, as well as various distractions and even the love of nature and its specifics.

At first glance, figuring out how the brain responds to active participation in brainstorming is a very difficult task.

You can't just stick a man in a CT scanner once he's started brainstorming.

But in fact the task is much simpler. When remembering real events, the brain activates the same patterns that were active during the real formation of the project concept, at least with accuracy to a certain task of the startup project, and it is precisely this accuracy that provides the experiment with the systematic transformation of information and technical requirements and characteristics in order to create a complete basic framework of the project with the introduction of active elements of artificial intelligence and artificial neural networks into the structure of management and control.

The participants of a relatively long period of the initial formation of the innovative concept of a startup project were simply puzzled to recall the state they experienced during the initial period and consistent intensification of the subsequent brainstorming steps with constant correction and refinement of the assignment to analyse the structure of the brainstorming with maximum approximation to the structures of the technique and technology of the startup project;

Strictly speaking, the results obtained in the intermediate stages of brainstorming correlate precisely with the sequential analysis of the achieved effect rather than with the actual performance of the technical characteristic, although brainstorming processes these states in a similar way;

The effectiveness of brainstorming manifests itself in many forms: from the emergence of an idea or interrelated ideas of a new invention to the anticipation of parameters of technical characteristics, the use of which creates a clear picture of the algorithm and structure of the claims in general to independent claims in particular, the combination of which creates, in conditions of complete non-obviousness, a clear relationship between the independent claims and the dependent claims attributed to them;

Such variants of processing and analysing the state of structure and schematic solutions of a new innovative product allow to create an integrative invention, which takes into account all the nuances of technical characteristics when including in the hierarchy of all the interrelated parameters, both hardware part and system part in conjunction with the applied software structures and ways and methods of their effective use;

Figure 0 - 1, is an example of the setting and combinations of brainstorming equipment in production structures of scooter manufacturing.

Figure 1, - the result of brainstorming - a robot with overhead control and monitoring systems with artificial intelligence elements and artificial neural networks.

Thus become real as a result of brainstorming of inventions with claims, - Apparatus (device), system (supersystem with input subsystems), programme (there are a lot of variants and executions here) and associated method (method);

The formation of so-called smart technologies in the brainstorming process opens up new possibilities, in which all parameters and features of their technical characteristics can be analysed in stages, with full control of changes in parameters at each stage of brainstorming;

For example, colleagues would say to a participant in the experiment: "Imagine you are seeing your project for the first time . How do you feel?" Did you achieve the ideal end result when you realised your idea?

Experts have found that the highest brain activity occurs precisely when experiencing the presence or absence of the effect of achieving the ideal end result .

"In this case, there was activation deep in the brainstorming system and the emergence of new quite effective brainstorming topics;

Brainstorming participants recorded dramatically increased brain activity during brainstorming associated with rooms whose design can be classified as nature-loving.

(That is, "love for birch trees" or other important landscape, judging by the results of the experiment, is not a myth).

The researchers found that a love of nature activated the brain's work organisation system, but areas related to social behaviour and non-project-related areas were not activated.

With pets, the situation looks a little different. "The love of pets may elicit in brainstorming additional brain activity related to sociality.

Analysing the physiological part of brainstorming may seem like a cold and unnecessary experiment, but experts and leading startup employees hope that their research and experiments can be used to better treat depression or solve problems in brainstorming organisation using 40 methods and techniques to achieve the ideal end result according to the theory of inventive problem solving.

Figure 2, - current cyber security options and choices.

2. Principle of judgement

• to separate the "interfering" part ("interfering" property) from the object;

• highlight the only part (the right property).

To solve problems of this kind in modern conditions it is necessary to analyse the usefulness and efficiency of parts and components, to identify parts and components that interfere with the implementation of the main working process of the supersystem under consideration and, within this supersystem, the nature of interaction of the incoming subsystems.

For such analysis in the conditions and taking into account the opportunities arising in today's conditions it is necessary to form a model of actions with application of elements of artificial intelligence and artificial neural networks, and also for performance of all control and measuring operations it is expedient to apply principles and devices on the basis of and in accordance with principles of electromagnetic resonance spectroscopy.

Figure 3, - current options and options for implementing elements of artificial intelligence and artificial neural networks in smart manufacturing.

For allocation of incoming subsystems and their degree of usefulness for the working cycle of supersystems, as well as for identification of the only necessary and most effective subsystem from all subsystems in today's conditions it is important to have or create a simulation programme with the help of which it is possible to carry out a step-by-step simulation of the working cycle of interconnected subsystems in the supersystem and in the system analysis to identify both the interfering subsystem and the only necessary part - subsystem;

After performing such a system analysis, it is possible to run the entire supersystem through the work cycle in a sequential manner and to classify the entire hierarchy of subsystems as meeting the characteristics of smart technologies;

Another function of the analysis is the necessary systematisation of the characteristics of subsystems, to determine the degree of utility and the level of utility to determine the only necessary part in the subsystem equivalent and its correspondence to the parameters and characteristics of the supersystem.

Figure 4 , - also current options and options for introducing elements of artificial intelligence and artificial neural networks into smart manufacturing.

3. Local quality principle
• to move from a homogeneous structure of the object (or external environment, external influence) to a heterogeneous one;

- different parts of the object should have (fulfil) different functions;
- each part of the facility should be in the conditions most favourable to its operation.

Naturally for the object should be created conditions most favourable for work including conditions for psychological rehabilitation of stress dependence, especially in brainstorming in the conditions of innovative variants of processes development.

Naturally different parts or components of the object should have and certainly fulfil different functions, but to change the principle of local quality by forming instead of a homogeneous structure of the object some heterogeneous background or composition can hardly help in forming an optimal object.

The widespread introduction of artificial intelligence elements and artificial neural networks into the control and monitoring systems makes it possible to manage and control all processes together with the fulfilment of the local quality principle.

In addition, systems based on electromagnetic resonance spectroscopy should be an intelligent tool for controlled performance control and the achievement of local quality principles.

4. The principle of asymmetry:
- to go from a symmetrical shape of the object to an asymmetrical one;
- if the object is asymmetric, increase the degree of asymmetry.

It is not always possible to neglect the symmetry of an object and go from symmetrical to asymmetrical form; All design techniques, all design programmes are aimed at maximising symmetry, not the other way round.

If some or some asymmetry takes place in an object, it is not clear how increased or even hypertrophied asymmetry can help in optimisation of the object or creation of a new object? How can this factor affect the operational and consumer properties of the object?

In this regard, the capabilities of artificial intelligence and artificial neural networks allow to predict the levels and radii of asymmetry and control the processes of influence of asymmetry parameters on the performance of the whole object, - all its supersystems and subsystems included in them.

In addition, dynamic asymmetry should be continuously monitored using non-contact real-time measurement techniques.

5. Principle of association:

- to connect objects that are homogeneous or intended for related operations;
- combine homogeneous or related operations over time.

If a facility has homogeneous components that duplicate each other's functions, combining them is unlikely to help achieve improved qualities and characteristics for the facility.

Combining homogeneous or related operations requires algorithm changes and reprogramming of the object processors, which is practically impossible in today's conditions.

To some extent good work in control and monitoring systems of elements of artificial intelligence and artificial neural networks allows to define necessary and possible character of the principle of association and besides application of principles of electromagnetic resonance spectroscopy as tools of control allows to increase sharply speed and accuracy of control that protects from accident.

6. The principle of universality:

- the object fulfils several different functions, thus eliminating the need for other objects.

In this situation, the question of the appropriateness of the object to fulfil functions inherent in other objects becomes obvious and legitimate.

Who needs it, who would go to the extent of combining the functions of other tools in one product, such as a tool.

Only one operator can use such a tool and if at the same time it is necessary to use a new function incorporated in the invention, it is necessary to have another such tool, i.e. commercially such an invention is of no use and will not be supported by investors.

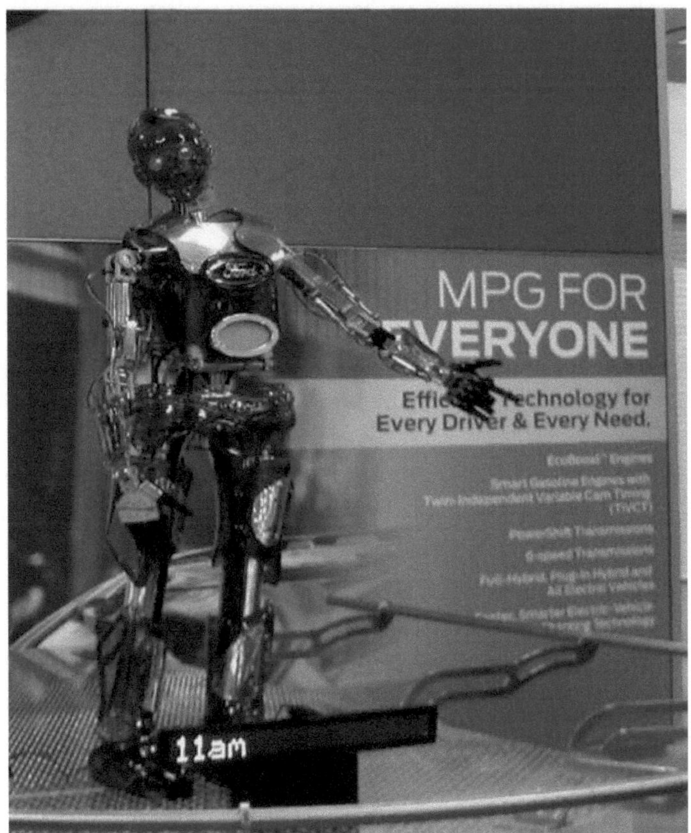

Figure 5, is an example of a robot used in the automotive industry, containing elements of artificial intelligence and artificial neural networks in its control and monitoring systems.

In addition, the use of programmes and mobile applications with control and monitoring parameters predetermined for each object, this kind of universality introduces the need for software coordination of output characteristics of each subsystem, but all this only within the supersystem and its control, analytical and management elements.

In addition to the above, it is also necessary to return to the issue of non-obviousness of the basic technical solution and determination of the level of its fundamental novelty.

7. Matryoshka doll principle
- one object is placed inside another, which in turn is inside a third, and

so on;

- one object passes through cavities in another object.

If a super object consists of incoming objects inside one another or one object passes through cavities in another object, then the conclusion about autonomy and independence of each of the objects is inevitable - and then it can be assumed that each of the objects is so autonomous and original that it is the subject of an independent invention, again in the presence of signs of obvious autonomous non-obviousness.

If this statement is viewed in terms of the interaction of the supersystem and its constituent subsystems with their hierarchy of software systems, this factor significantly narrows the functional independence and autonomy of each object.

In this case the possibilities of management and control closely depend on the speed of processors of subsystems and as a consequence the penetration of incoming objects - subsystems in the central head objects - supersystems are significantly limited and require special software control with high speed, which is not always economically feasible.

It is also necessary to take into account the proven fact that the real performance of for example sensors working on the principles of electromagnetic resonance spectroscopy is one control or measuring operation in 10 milliseconds.

It is also necessary to note the manufacturability and reliability of the structures having an internal structure equivalent to the matryoshka doll principle.

That is, this principle allows to simplify considerably both the design and operation of units and mechanisms, while maintaining the quality indicators of all levels.

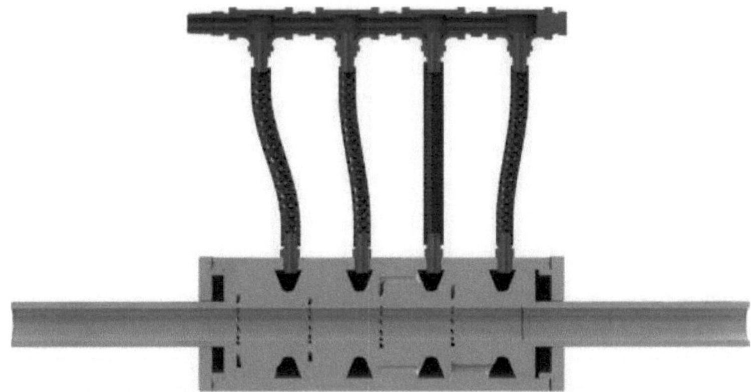

Figure 6, - the figure shows a cross section of a device with 4 vortex generators , designed to form a vortex tube in a gaseous fuel flow.

All tubular structures in this device are coaxial and generally function as the matryoshka doll principle.
There are many such variants in technology and all of them are used in many cases, especially in fuel systems for gaseous fuels.
8. The Anti-Weight Principle:
• compensate for the weight of an object by coupling it with another object that has a lifting force;
• compensate the weight of the object by interaction with the medium (due to aerodynamic - and hydrodynamic forces).
Balancing and equilibration are among the most important techniques for calibrating and adjusting innovative objects ; All modern versions of computer design programs do this automatically, so there is little point in helping them manually.
Implementation of smart technologies is usually carried out within the framework when all subsystems in interaction with the supersystem have a complete version of mutual balancing and equilibrium both on the systems of vertical integration and on the systems of horizontal integration, and the systems of balancing of mass and weight parameters should have a clear balancing software algorithm under the control of system elements of artificial intelligence and artificial neural networks.
9. The principle of prior anti - action
• give the object stresses opposite to unacceptable or undesirable operating stresses in advance;

• If by the conditions of the task it is necessary to perform an action, it is necessary to perform an anti-action beforehand.

In the modern design process, all preliminary evaluation and calculation operations for optimising strength, temperature and other parameters are carried out using design software, i.e. such optimisation is carried out quickly and optimally by the computer.

Besides all the load on elements of subsystems design in dynamics of combination of actions and anti-actions between subsystems of the supersystem in today's conditions as a rule have a clear and unambiguous programme code with binding and interaction with the central processor of the supersystem.

10. Pre-action principle:

• perform the required action (fully or at least partially) in advance;

• arrange the objects in advance so that they can come into action without time-consuming delivery and from the most convenient location.

In computer-aided design almost in all such programmes of this kind and purpose there is a possibility of stimulation and simulation of the working cycle of the object, so that as a rule such operations necessarily include all possible variants and versions of actions and functions of the object.

In addition, programmes and algorithms take into account the dynamics of processes and actions and coordinate these elements of the programme and dynamics with the central processor of the supersystem.

11. The principle of the pre-padded cushion

- compensate for the relatively low reliability of the facility with pre-prepared emergency means.

From the point of view of modern relations between investors and inventors, it is clear that a low reliability of the object will reliably and firmly close the way to the market.

No compensation actually compensates for anything, only a reliable and absolutely workable design determines the real commercial success of a facility.

Thus, as a rule, the software of the supersystem takes into account all issues and factors of sources of reliability level reduction and provides dynamic solutions for reliability level strengthening.

In addition, today's additional opportunities to increase the reliability of objects are also provided by the use of composite materials, the strength and reliability of which significantly exceed the capabilities of traditional materials.

12. Principle of equipotential coupling

- change the working conditions so that you do not have to lift or lower the object.

If the principles of operation of the object provide for static operating conditions, then of course it makes no sense to raise or lower the same object or vice versa, if the object operates in a developed or local dynamic mode, changes in the operating conditions of the object will not only not lead to optimisation of its operating cycle, but will generally make the operation of the object impossible.

As a rule, in today's smart technology environment, all work movements are strictly limited and speeds are strictly controlled online in real time using control devices based on the principles of electromagnetic resonance spectroscopy.

13. Reverse principle

• Instead of the action dictated by the task conditions, perform the opposite action;

• to make the moving part of an object or environment stationary and the stationary part moving;

• turn the object upside down, twist it.

If an object has a basic algorithm for its working cycle and the system of control and monitoring of the working cycle of this object is encoded in the software processor, then the reverse action requires a complete change of the algorithm and, accordingly, reprogramming of the processor, which is practically impossible neither for financial reasons nor under the conditions of fierce competition in the circumstances of globalisation.

In this regard, the design and programming process takes into account the dynamics of actions and movements and its relationship to the dynamics of all subsystems within the dimensions and surface quality of the supersystem.

Thus, this principle cannot directly solve design problems and should be a set of coordinated movements involving sequential movements of subsystem design elements while maintaining the required level of mutual balance and dynamic tasks to be performed by the subsystem elements.

14. The principle of the spheroidal concept

- move from rectilinear parts to curvilinear parts from flat surfaces to spherical surfaces, from cube and parallelepiped parts to spherical structures;
- use rollers, balls, spirals.
- to change from rectilinear to rotational motion, to use centrifugal force.

If we consider the transition from rectilinear parts to spherical and curved parts from the point of view of manufacturing, even with the use of the latest machining methods on digitally controlled machining centres, it is clear to any specialist that we are talking about an order of magnitude more expensive process.

Also, the transition from rectilinear to rotational motion leads to additional kinematic problems, which also systematically deteriorates the design and performance properties of the product

In addition, similar problems can be successfully solved today with the use of composite materials having a capsule structure and consisting of spherical nano-capsules, each nano-capsule having a core made of artificial diamond with a shell made of a ductile metal, most often copper

Such products are manufactured in a mould, which uses the cold flow principle to form the outer shape.

15. The principle of dynamism
- the characteristics of the object (or external environment) must change so as to be optimal at each stage of operation;
- to divide an object into parts that can move relative to each other;
- if an object is generally stationary, make it mobile, moving.

The most successful objects, based on the concepts of modern technology, are those in which the working cycle is carried out without moving parts or elements

Naturally, any object must be optimal in all modes of operation; If we apply the technique of dividing the object into parts capable of moving relative to each other, it is necessary to ensure the optimality of each such part, keeping the logic of the above-mentioned.

Not to mention that ensuring local optimality is a very expensive and not always possible in such a situation, the creation of autonomous optimal components or parts capable of moving relative to each other and independent of each other is the creation of each part as a separate independent and optimal invention, which is contrary to existing patent regulations.

16. Principle of partial or redundant action

- If it is difficult to get 100% of the required effect, it is necessary to get "a little less" or "a little more" - the task will be considerably simplified.

The question is why simplify the inventor's task by reducing the percentage of the required effect? How will investors react to this reduction?

When considering modern inventive problems, it is difficult to agree that anyone knows exactly what 100% of the required effect is ?

How will this knowledge affect the commercial sector of the invention implementation, let's say we simplify the solution of the problem for the inventor, but if as a result the simplified solution will not meet all the requirements of the market, is there any point in all these simplifications ?

At the time of development of the theory of inventive problem solving, at least in the Soviet Union, commercial issues were of secondary importance and this determined the attitude to the real effectiveness of technical ideas fulfilled at the level of invention.

Knowing what can be considered as an effect that will ensure commercial success, knowing what can be considered as 100% of the required effect, can in itself ensure commercial success when performed.

17. The principle of going to another dimension:

• difficulties related to the movement (or placement) of an object along a line are eliminated if the object acquires the ability to move in two dimensions (i.e., in a plane). Correspondingly, the problems associated with the movement (or placement) of objects in one plane are eliminated when moving to space in three dimensions;

• use a multi-storey layout of facilities instead of a single-storey layout;

• tilt the object or put it "on its side";

• to use the reverse side of this square;

• use optical flows falling on a neighbouring area or the reverse side of an existing area.

This principle somehow bypasses the creation of absolutely new, as they used to say - pioneering inventions; according to the methods that implement the transition to another dimension, set out in the theory of inventive problem solving, today at best it is possible to improve and optimise the existing technical solution;

As a rule, protection of such improvements as intellectual property objects is extremely difficult and it is generally very difficult to divide this technical solution into features related to the object before the improvement and acquired by the object in the process and after the improvement.

In addition, the software part of the product should not be overlooked, which is expressed by the introduction in the claims of all the necessary interrelationships between the subsystems and the chain of interrelationships between the supersystems, which should be expressed in the claims as apparatus, programme, system and associated method, as well as the software and algorithmic interrelationship between all the subsystems, but without contradicting the supersystems.

18. The use of mechanical vibrations:
- to set the object in oscillating motion;
- If such a movement is already in progress, increase its frequency (up to ultrasonic);
- use the resonant frequency
- to use piez vibrators instead of mechanical vibrators;
- use ultrasonic vibrations in combination with electromagnetic fields.

Modern innovative technical solutions, which form the basis of many innovative products, apply system mechanical oscillating circuits everywhere to solve the most complex technical problems, especially in non-contact metrology.

As it turned out, the highest efficiency and accuracy is shown by magnetic resonance combined sensors, as well as by the

piezoelectric motors, which in the latest developments have shown excellent capability to operate as stepper motors.

Thus, this method and technique is still fully relevant today and has also shown excellent adaptability to the latest processor control and monitoring systems and to the applicability of composite materials.

19. Periodic operation principle
- switch from continuous to periodic (pulsed) operation;
- if the action is already performed periodically, change the frequency;
- use the pauses between pulses for another action.

One modern technique is to apply pulse technology to solve very important problems especially in laser control systems and RF drivers.

The technique of using the pause between pulses is also widely used in different variations, so that this technique is also still valid for the most important and horizontally integrated technical solutions.

20. The principle of continuity of utility

• operate continuously (all parts of the facility must be operating at full load at all times);

• operate continuously (all parts of the facility must operate at full load and at controlled dynamics at all times);

To fully understand the importance of this technique, it is necessary to note the subtleties of the modern processor control process in which the pause in the working cycle of any complex is also an object for control.

That is, continuity of work includes all parts of the work cycle, including pauses.

This technique explains the necessity and feasibility of including an algorithm as part of a modern invention.

In addition, switching of the direction of actions and movements, as well as changing the dynamics of pulsations require programme control switching of the direction of movement of control elements and tools, of which the most effective have shown themselves to be non-contact electromagnetic resonance tools possessing in the case of the use of a flat coil extremely high speed.

In addition, the use of a multilayer flat coil allows you to dramatically increase the accuracy and depth of penetration of the signal pulse and sharply reduce the dependence of measurement accuracy on electronic noise.

20. The principle of slippage

- Conduct the process or individual steps (e.g. harmful or hazardous) at high speed.

This technique is a basic part of modern integrative inventions, in which the structure of novelty and operability is based on a programme, an algorithm and a method implementing their integrated combination.

21. The principle of "turning harm into good"

• utilise harmful factors (in particular, environmental hazards) to produce positive effects;

• eliminate the harmful factor by compounding it with other harmful factors;

• amplify the harmful factor to the point where it is no longer harmful.

This technique seems to me to be the most important for understanding modern invention, which can be the basic foundation of a commercially successful product.

For example, accelerated electrode degradation in an electrochemical reactor, which is catastrophic in terms of electrolyte contamination in the galvanic process, is a factor in the success and significant performance improvement of the electrocoagulation process used in industrial wastewater treatment and regeneration systems.

Application and combination of all components of this principle allows to form new products and technological principles, especially in electrochemistry and waste-free production in any branches of industry and agriculture.

23. Feedback principle
- provide feedback
- if there is feedback, change it.

The principles of feedback are used today in almost every processor-controlled product, but all feedback components are familiar components of the product and for the inventor and in general any designer the application of feedback systems is reduced to a simple selection of components.

24. Intermediary principle
- Use an intermediate object that carries or transfers the action;
- temporarily attach another (easily removable) object to the object.

The principle of the intermediary or technological witness is still widely used today, but its application has no effect on the commercial results of a product that was manufactured using this technique.

All this can be attributed to the peculiarities of the process, and patenting of the process without linking it to the product is almost never practised today - everyone is interested in the product itself, and how it was produced with or without the use of the intermediary principle is of little concern.

25. Self-service principle
- the facility should be self-sustaining with auxiliary and repair operations;
- utilisation of waste (energy, substance)

This principle is now the most important of the 40 techniques inherited from the classical theory of inventive problem solving.

It can be said that this technique consists of two important techniques - the formation of waste-free production and self-sufficient autonomous innovative products.

Waste-free production is a special issue and its scope requires a special study and description, but fundamentally the application of algorithm methods in which the waste is a part or component of the initial fuel, for example, it is as they say the highest pilotage of the innovation process.

It is enough to give only one example, - condensation of water from exhaust gases and use of this water in the process of fuel emulsion formation.

26. Principle of copying:

• Use simplified and cheaper copies of an inaccessible, complex, expensive, inconvenient or fragile object instead of an inaccessible, complex, expensive, inconvenient or fragile object;

• replace an object or a system of objects with their optical copies (images). Use in this case change of scale (increase or decrease copies);

• if visible optical copies are used, switch to infrared and ultraviolet copies

Today such recommendations from the technical point of view look like obvious and outright stupidity, but even if we consider all this from the point of view of patent law, copying in such a scenario will never lead to the creation of an original technical solution with world novelty.

27. Cheap non-durability in lieu of expensive durability

- replace an expensive object with a set of cheap objects, while sacrificing some qualities (e.g. durability).

Such a technique can only be justified if it is a single-use product.

Any cheapening of the product, realised at the expense of quality, cannot be understood by the market and therefore cannot be accepted, i.e. the consequence of the use of such a technique can be a complete commercial failure.

28. Replacement of the mechanical system

• replace mechanical circuitry with optical, acoustic or "smell" circuitry;

• use electric, magnetic and electromagnetic fields to interact with an object;

• to move from fixed to moving fields, from fixed to time-varying fields, from unstructured to structured fields;

• use fields in combination with ferromagnetic particles.

Everyone knows that the most reliable objects are those that have no moving parts.

Therefore, any technical equivalent of the moving parts of a product that can fully replace the moving parts with fixed parts or that operates on a principle that performs the same functions as the moving parts can help to create an innovative object in which, due to the absence of moving parts, reliability, durability and performance will reach a level that will ensure the product's competitive success and success in commercialisation.

29. Use of pneumatic and hydraulic structures

• use gaseous and liquid parts instead of solid parts of the object;

• use electric, magnetic and electromagnetic fields to interact with an object: inflatable and hydro filled, air cushion, hydrostatic and hydro reactive.

The execution of this technique is extremely difficult from the technical point of view; For commercialisation of this technical solution it is very difficult to find arguments for the investor and to explain that, for example, in a traditionally accepted and understandable design of any product, it is necessary to apply an exotic solution or solutions associated with commercially not always reasonable replacement of the basic principles of the product.

Such a replacement is always associated with a significant increase in the cost of the product, which as an argument always works against such changes.

The analysis of formulations of the principle in view of application for development and design of new design programmes, new high-speed computers of a traditional type and a prospect for the future with possibility of application of quantum computers and their special design modifications.

Figure 7, - the figure shows the current interpretation of the application of gas composite supply and control to the combustion chamber of an industrial boiler with independent control of each branch of the supply pipework.

Figure 7 - 1 , - the figure shows a gas composite supply system using an intermediate receiver ; The same configuration can be used for the hydraulic version of diesel fuel supply and natural gas dissolution.

The same configuration can also be used to form a fuel composite comprising diesel fuel and compressed air or an oxidising equivalent thereof.

30. Use of flexible shells and thin films
- use flexible shells and thin films instead of conventional structures;
- isolate the object from the external environment using flexible shells and thin films.

It is useful to divide this technique into micro and macro components.

This technique is also an important element for e.g. nanoscale technologies.

Take for example the process of preparing an emulsion from fuel; If an emulsion is formed, for example, from diesel fuel and water, the only guarantee for the success of such a fuel emulsion is its structure, in which micro-droplets of water are surrounded by a thin film of diesel fuel.

After injection of such an emulsion into the combustion chamber, the water, which evaporates before the diesel fuel, breaks the diesel fuel shells into nanoscale particles during evaporation and this is much more effective for an optimal combustion process than, for example, a sharp increase in injection pressure, which entails the need to strengthen the structural elements of the cylindrical piston group of the engine.

The same technique can be the basis for inventions in which, for example, the electronic board is covered with an ultra-thin film of synthetic diamond in places intended for intensive heat exchange.

31. Application of porous materials

• make the object porous or use additional porous elements (inserts, coatings, etc.);

• if the object is already porous, pre-fill the pores with some substance.

To begin with, the technical and technological method of using porous structures of an innovative object is widely used in order to give the object the necessary properties.

Very often in modern developments there is a need to form pseudo-porous materials, which after obtaining the properties of pseudo-porosity acquire the qualities of a composite material.

A porous material does not always have to be hard.

It can also be liquid, e.g. as an emulsion, like water to oil.

In such an emulsion, water droplets are introduced into gravitational breaks in the flow of oil or, for example, diesel fuel, i.e. the cavities or pores are filled with some substance - water.

Such materials are known and quite widely used in the form of a viscose fabric coated in vacuum with carbon; such a fabric is a

is a good conductor of electricity and can withstand operating temperatures of 4000 degrees centigrade.

32. Principle of colour change

• change the colouring of an object or the external environment;

• change the degree of transparency of an object or the external environment.

If we extend the application of this technique to optics and optical memory, then an explanation can be found for its application.

Under US patent law, for example, the use of this technique is in principle contrary to the principle of non-obviousness of the technical solution underlying the invention.

This statement may be acceptable up to the moments related to premises and equipment for startups and for timely compensation or rehabilitation of deviations in the psychological climate, especially in brainstorming.

33. Principle of homogeneity

- objects interacting with this object must be made of the same material (or close to it in properties).

This rule or technique is quite controversial and it is not easy to follow when designing a project.

Besides, it is not always clear - what does the design as a whole gain by blindly following this rule?

34. Principle of discarding and regeneration of parts

- The part of the object that has fulfilled its purpose or has become unnecessary must be discarded (dissolved, vaporised, etc.) or modified directly in the course of work;

- expendable parts of the facility must be recovered directly from the work.

This technique can be referred to the means of formation of waste-free production.

Environmental safety is now an extremely important aspect of any production process and the above technique at least does not contradict the general logic of the construction of this production process.

However, this principle is not ideally suited to all projects and many processes, especially in the production of, for example, especially pure materials from hen eggs (especially pure egg oil; alpha lecithin ;lysozyme; albumin; multivitamins) require absolutely specific production conditions.

35. Change of physical - chemical parameters of the object

- change the aggregate state of the object
- change the concentration or consistency
- change the degree of flexibility
- change the temperature

This technique is widely used and its area of use is constantly gaining new segments

36. Application of phase transitions

- Use phenomena arising from phase transitions, e.g. volume change, release or absorption of heat, etc.

In modern industrial systems, especially thermal - dynamic systems, phase transitions are an integral part of the technological hierarchy.

As an example, pre-mixing of fuel with air, whereby the injection volume is significantly increased and, at the same time, the temperature of the gas phase of the fuel mixture is lowered during injection and expansion.

In addition, an example can be given of the production of so-called energy fuels, in which combustible gases are dissolved in liquid fuels in developed aerodynamic and hydrodynamic flows.

37. Application of thermal expansion:

* utilise the thermal expansion (or contraction) of materials;
* use several materials with different coefficients of thermal expansion.

This technique is the most applicable in modern measurement technology and is quite relevant

The analysis of formulations of the principle in view of application for development and design of new design programmes, new high-speed computers of a traditional type and a prospect for the future with possibility of application of quantum computers and their special design modifications.

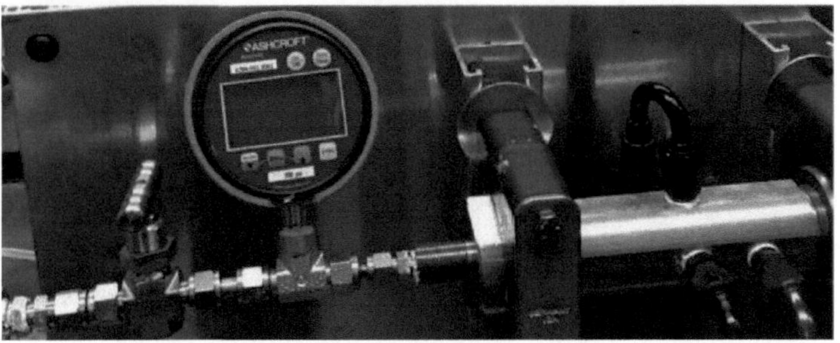

Figure 8, - the figure shows the principle of device installation, which allows to compensate and neutralise the phenomena of thermal expansion of structural materials of device parts forming the cross-sections of working channels.

This factor is particularly important if the internal structure of the object contains, for example, working channels with micron cross-sections.

When changing the operating temperature in the body of an object with microscopic channels, a temperature change of even a few degrees can lead to blockage of the channels due to thermal expansion of the materials.

Figure 8 -1, - the figure shows the fuel mixture preparation system with a compensating block compensating for thermal expansion and changes in the cross-sectional area of the working channels of the device.

38. The use of strong oxidising agents:
- replace ordinary air with enriched air
- replace enriched air with oxygen
- ozonise oxygen
- replace ozonated oxygen (or ionised oxygen) with ozone.

This technique also remains relevant and is widely used in various fields of engineering and technology.

39. Application of inert medium:
- replace the normal medium with an inert medium
- vacuum

This technique is extremely relevant today for the semiconductor and microelectronics industries.

For better results in micro-miniaturisation, all processes are carried out in clean rooms, in an inert environment or in vacuum.

That is, for any invention in these advanced fields, an inert environment or vacuum chambers are a mandatory background that provides the required level of quality.

40. Applications of composite materials:

- move from homogeneous materials to composite materials.

As is well known, the substitution of one material for another is not recognised as an invention; the composite material itself is the subject or object of the original invention, but very often, when an ordinary construction material is changed to a composite, the properties and capabilities of the product are so changed that the product in which the composites are used becomes a completely new, unknown before, with completely new functions and unusual technical characteristics.

Of course, in order to make such a replacement it is necessary to perform such a volume of work, which is comparable to the development of a fundamentally new technology or a fundamentally new product, and this is possible only for companies with powerful research departments.

Integration of different technological directions in any modern innovative project, very often requires the use of composite materials of absolutely new properties and often in absolutely unusual quality in order to obtain an ideal final result.

These new composite materials may belong to secondary integration levels, for example, if a new type of internal combustion engine is invented, a composite fuel in the form of a mixture consisting of, for example, petrol and ethanol or an emulsion consisting of, for example, diesel fuel and water may be used in the same engine design instead of the traditionally accepted fuels in order to obtain an ideal end result, for example, in the field of reducing the concentration of soot in exhaust gases.

CHAPTER 2

Law of completeness of parts of the system

A prerequisite for the fundamental viability of a technical system is the presence and minimum operability of the main parts of the system.

Law of energy conductivity of the system

A prerequisite for the fundamental viability of a technical system is the through passage of energy through all parts of the system.

The law of harmonisation of the rhythm of the parts of the system

A necessary condition for the fundamental viability of a technical system is the coordination of rhythmics (frequency of oscillations, periodicity) of all parts of the system.

The law of increasing the degree of ideality of a system

The development of all systems is in the direction of increasing degrees of ideality.

Law of uneven development of parts of the system

The development of parts of the system is uneven. The more complex the system is, the more uneven are the processes of development of its parts.

Law of transition to the supersystem

Having exhausted the possibilities of development, the system is included in the supersystem as one of its parts. In this case, further development takes place at the level of the supersystem.

The law of transition from the macro to the micro level

The development of the working bodies of the system proceeds first at the macro level and then at the micro level.

The law of increasing the degree of real-field bonds

The development of technical systems goes in the direction of increasing the number of substance-field connections.

TRIZ attempted to formulate the laws of technical systems development, which should have formed the basis of TRIZ and the general methodology of problem solving. However, most of the formulated laws are not such laws. Rather, they should be called laws of technical development, which are far from being complete. For this reason, a coherent methodology of problem solving based on the laws of development never appeared. And the formulated laws were mainly used as methodological justifications for the given examples of inventions. The factors of commercial expediency were completely excluded and their influence on the methods of using the constantly modifying laws of development of technical systems for such modernisation, optimisation and modification of the object that can evolutionarily turn the invention into a demanded innovative product was not taken into account.

The recent patent disputes between the largest technological companies of the world show and prove that the laws of technical systems development formulated in TRIZ cannot reflect the whole variety of tasks, functions and features of a modern multifunctional object, and taking into account all the new and emerging factors characterising an innovative object, it is necessary to redefine these laws by linking them to the laws of development of commercial structures and commercialisation of innovative ideas.

The dialectical approach (analysis of contradictions) embedded in the main tool of problem solving, which was ARIZ, was distorted by the introduction of new concepts (technical and physical contradiction). These new concepts distorted the essence of the dialectical contradiction formulated in dialectical logic, which led to difficulties in identifying the contradiction when trying to solve real inventive problems using ARIZ. We should focus on this separately - what can be considered a real inventive problem? How can a correct or erroneous formulation of an inventive problem affect the commercialisation of the resulting invention? Is it possible to protect the resulting technical solution from unauthorised copying? The search for answers to all these and many other questions is now becoming a major part of the dialectic of creating a strategy for patenting and licensing inventions.

Improvement of ARIZ (creation of new modifications from ARIZ-77 to ARIZ-85B) followed the path of complicating the algorithm rather than eliminating inaccuracies in the contradiction detection procedures. As a result, the last official modification of the ARIZ-85B algorithm turned into an extremely cumbersome design that is of little practical use.

1. TRIZ has never found clear mechanisms of transition from a formulated contradiction to its practical resolution. This created serious difficulties in solving real-world problems with the help of TRIZ.

2. TRIZ declared the rejection of the methodology of activating the search of variants, but the main part of the so-called TRIZ tools represented exactly such methods (the method of little men, the RBC operator, real

field analysis).

3. Substantial field analysis was presented in TRIZ as a scientific approach based on the analysis of regularities of structural development of technical objects. However, the assumption of the use of non-existent physical fields in real fields, as well as the possibility of ambiguous interpretation of real-field constructions and rules of their transformation, rather allow us to refer real-field analysis to the methods of activating the search of variants, but not to scientific analysis.

4. The closest to the idea of formalising the procedure of inventive problem solving was the creation in TRIZ of tables and techniques for resolving technical contradictions. This approach was based on the statistical analysis of the then existing invention descriptions. However, despite the existing prospects, it was not further developed in TRIZ, and due to a number of shortcomings and obsolescence of statistical conclusions, it lost its relevance for practical use.

5. There is a widespread illusion about the possibility of TRIZ implementation in real production. In its essence TRIZ is an individual method of problem solving, the application of which is a personal choice for a person. For this reason, it is impossible to make TRIZ a part of a particular production process. At best, an enterprise can organise TRIZ training for its employees in order to enhance their creativity.

All inventors know that sometimes technical solutions are created which in real conditions act, work and solve a lot of problems, which already at the setting pushed the inventor to innovative analysis and initiated his purposeful creative activity, and there are unpromising technical solutions, which are created in isolation from the real reality and solve absolutely nothing, except for the realisation of ambitious claims to any (most often useless) - but an idea in the field of engineering and technology;

Besides, technical solutions arising in any local area, necessarily directly or indirectly affect the existing stereotypes and psychological barriers arising on their basis, preventing from overcoming technical-technological contradictions arising on the basis and in the development of these psychological barriers.

Twenty years ago, the need for inventions of the second group and the equally important need to take into account the influence of psychological barriers was somehow justified by their auxiliary role as a basis for selective selection of the most effective technical solutions.

The emergence of information technologies and a sharp reduction in the time cycle for the development and transformation of an inventive idea into a product that is actually necessary, demanded by the market and realisable, the increasing complexity of the technical and technological components of new products, causing a proportional increase in the cost of manufacturing prototypes of the invented product and their testing, make us consider the possibility of creating technical solutions with auxiliary innovative functions in a completely new way.

Now, if an inventor wants his innovative ideas to be used, he should be more universal and should possess not only foresight, intuition and to a certain extent developed imagination, but also be a practically universal specialist, at least feeling and (better if) well understanding the commercial and consumer demands of the market, regardless of stereotypes and psychological barriers combined with them.

There are several basic directions, which in the current conditions have a decisive influence on the fate of new ideas and, taking into account which, may allow to ensure a real and high level of commercial success, or, neglecting which, will forever close the way for the idea to be realised in any commercial form.

I propose to review some of these basic directions (of course, the scope of the article allows to do it only in the thesis form). :

Availability of fundamentally new materials

Let's consider for example two new technological directions, - creation of effective light sources on the basis of radiation of blue lasers (laser diodes) and creation of composite food products on the basis of effective systems of mixing and hydrodynamic activation.

Both mentioned technological directions for development require structural materials, which due to their properties will allow to obtain parameters in each direction, which are impossible to obtain by using conventional materials

The feasibility of new materials is not absolute and a primary compositional solution is required to enable their application.

For example, lasers require a highly efficient heat sink and current pulse dissipation system, and for its implementation it is necessary to make changes in the design of the laser housing that would make it possible to use a composite diamond-copper material, which by its parameters and properties is able to perform these functions

But it turns out that replacement of the material does not solve all the complex problems facing the creators of new lasers; what is the solution?, - the solution may be a compositional scheme, when, for example, in addition to the material in the system of technical composition is introduced a radio-frequency driver, which, in combination with a new material that performs the functions of cooling and dissipation of excessive heat and current pulsations, allows the laser to provide a pumping mode, which in turn allows for a laser with a power of 1 watt, to introduce a pumping current in the, say

Availability of composite materials

Information about scientific and technological developments in composites is eagerly awaited in the market and new composite materials are constantly appearing and their availability makes adjustments in the design of new products and in the technological process of their manufacture.

To link these new developments with the vast scientific, technical and technological experience gained in research and development enterprises, as an example, we consider the technique of creating nanoscale composites, the main component of which is nanoscale synthetic diamond powder, by the way, the technology of production of which was born in the walls of the Ukrainian Institute of Superhard Materials.

A significant novelty added to the above technology is the technique of high-speed nanoscale nanocoatings on nanoscale diamond powders, which are made of the most ductile metals such as copper, silver, gold, platinum and with further introduction of these components into a complex technological composition, with the introduction into the process of operation of initiation of cold flow properties and subsequent plastic calibration under ultrahigh pressure of the nanoscale capsules thus obtained; Thus, we see in this composite solution a consistent horizontally oriented integration of the technique of precision nanoscale coatings and the technique of cold flow initiation in the material coated with the same coatings.

Thus, in this publication the aim is to show that the available level of research and development in the field of composite materials science, allows, with the correct formulation of the problem to the developers of one component of the composition and, with in-depth cooperation with the developers of similar technologies, the product of which is another component of the composition, to create actual technologies and materials in demand on the market.

It is now possible to trace how the process of composing a composition of technologies and materials evolves:

Since the creation of complex integrated, energy-rich semiconductor devices, especially semiconductor lasers (laser diodes), the problem of efficient heat dissipation, heat dissipation, dissipation of current pulses and fluctuations has arisen.

The reason for this problem was the lack of structural materials, alloys and all kinds of combinations and combinations of materials, the ability to reliably and sustainably perform these functions.

All materials and their derivatives to one degree or another did not satisfy developers and operators, and only with the advent of the possibility of creating complex composite technical solutions, the problems of this level can be solved.

New directions in the technology of manufacturing elements

In composite technical solutions, the methods and technologies of manufacturing parts and components of the product, are the most important factor that determines the reality of the implementation of this product at all.

As an example, consider the injector or fuel injector of an internal combustion engine ; It is one of the most mass-produced products, with more than a billion such injectors produced worldwide per year.

For such a product its consumer value is determined by several key factors, - the diameter of the outlet holes and ensuring tightness at high fuel pressures (up to 2000 atmospheres); The execution of holes using conventional technology determines the limits of the minimum diameter of holes, and since at high pressures require holes whose diameter is measured in microns, the technology of execution of these holes should be, for example, laser; In this case, the inventor of a new injector must provide for a composite component, but it is not necessary to provide for the minimum diameter of the holes.

New consumer standards

Constantly changing conditions and demands of consumers create in combination with local consumption standards, with cultural and national traditions, on the basis of which local consumption standards, informal but tacitly present conditional consumer standards arise.

If the inventor's objective is to ensure the commercial success of his invention, then a basic understanding of the current consumer standard criteria, technical, operational and functional characteristics of the new product should be an essential part of his commercial strategy.

If the technological risk is eliminated because the new product has been tested and tested positively, the risk of commercial failure for this technologically quite successful product really remains if the inventors and their commercialisation partners have not considered or understood the nature of the consumer standard for their product.

New environmental standards

It is well known that the restrictive requirements of environmental standards are becoming more and more stringent; When working on the invention of a new product, the use of which in any way affects the limits of permissible parameters regulated by environmental protection standards, it is necessary to provide for full compliance with the requirements and restrictions of the applicable standards in the use of the new product.

As a rule, current standards are constantly being improved and requirements that are in force this year will routinely be tightened in a few years.

This is especially important for various power equipment and internal combustion engines.

There are cases when an internal combustion engine with an innovative cylinder layout and an extremely efficient fuel-saving system was invented, but the level of nitrogen oxides and soot in its exhaust gases exceeded the permissible levels under the future standard, which is due to come into force in two years.

This was enough to send the latest engine for further development, during which its technical solution was changed to the level of composite engine with the inclusion of an innovative system of fuel preparation and activation integrated with previous technical solutions, which allowed to reduce the concentration of harmful substances in exhaust gases.

The influence of fashion

Even for high-tech products there is fashion, as a kind of integral complex stereotype developed by time and practice, combining both objective and subjective commercial and operational factors.

The subjectivity of this concept is not always explainable by methods of direct logic, but it should be taken into account by the inventor both when developing a new product and when preparing its presentation for potential partners and consumers.

Very often, an innovative product contains elements, factors or features that future consumers are expecting and are favourably impressed by, but the authors concentrate on purely technical aspects and offer them something other than what they want to hear and see.

Availability and continuous intensive development of software products

Significant complication of technology and especially of various types of electronic and microelectronic devices has fundamentally changed the principles of their protection as objects of complex, multidimensional and multifunctional intellectual property.

For such objects essential differences concerning purely constructive features, circuit solutions, combinations of these solutions do not determine all aspects of the invention, because today very often all the listed features and differences can be realised in a working system or prototype only under certain conditions and possibilities of manufacturing and control technology, and very often it is the manufacturing that determines the main properties of the invention.

The development of processor control systems also determines the viability of the technical solution, which means that the algorithm or algorithms, programmes, feedback between the elements of the design or circuitry are becoming or have already become an organic part of the technical solution that forms the basis of the claimed invention.

Thus, it is necessary to combine or integrate several different technologies in a single description and this combination, the identified possible channels and links of such integration, should be presented in the claims in such a way as to prevent the patent office examiner from doubting the unity of all integrated distinctive features of the future invention and to divide it into a number of local technical solutions based on one technological direction.

Patenting opportunities in the field of information technology are a huge area of activity in modern society; and while not so long ago it was possible to characterise or restrict a particular technological sector in a very precise way, with the advent of high technology and its offshoot, information technology, such classification opportunities and protection mechanisms have changed significantly and have been transformed into a new system of technical, commercial and legal relationships.

In almost all processes, even relatively simple ones, their structure becomes integrative and includes technological methods, techniques and systems that have never been used before, and, in addition, the integration of classical technical solutions with new possibilities provided by information technology fundamentally changes the very concept of invention.

This factor, arising at the junctions of technologies, essentially changes the attitude to the formulation and protection of those elements and their combinations, which in such new conditions can be qualified as integrative technical solutions, corresponding to the main features of the invention and based on composite design and technological elements.

Turning an abstract idea into a real one as the most important part of the modern innovation process

The history of cross-patent suits in recent years between high-tech companies makes it necessary to check and imagine once again the common process of large companies to create inventions for which these companies then seek a patent.

So, who today in such companies invent, so to speak, out of industrial necessity or as part of their direct production duties ?

As a rule, all production units that make the product are outside the company and are located, at best, at the enterprises of related companies or, which is unfortunately typical today, in China.

That is, in a company, the duty to invent is assigned mainly to programmers, analysts of different levels and capabilities, but not to mechanical engineers or electromechanical engineers, who are not in the staff of these companies, because under the current structure of companies, they are not needed - they are needed in companies where production is concentrated.

That is, we can say that ideas of new products are successfully generated, but not in the form of a technical solution, but in the form of an abstract algorithm, a mathematical model, finally a programme, and all this has a very distant relation to the classical technical solutions, which can become the basis for a future invention.

Commerce cannot stand still and naturally thousands of patent applications are filed that have nothing to do with the invention, as neither explicitly nor implicitly they contain anything comparable to a full-fledged technical solution.

The vagueness of definitions and the almost complete absence of explicit cause-and-effect relationships, incompleteness of solutions lead to the fact that such incomplete solutions are perceived as abstract.

Not only that, but since most of the cases involve mobile communication tools and methods and tablet computers, it is very difficult, if not impossible, to distinguish one from the other.

It seems to us that a clear and unbiased patent law, which does not allow for compromise, should play a major role here.

As practice shows, loopholes and inaccuracies in legislation, compromises with non-mechanical and non-electromechanical basic content of inventions should be excluded by legislation.

Today, unfortunately, most of the inventors (in quotes) are lawyers who see a patent as a means of putting pressure on competitors.

Google, Facebook and six other companies have spoken out against IT patents that describe "abstract ideas," TechCrunch reports.

According to the companies, such patents pose a threat to technological progress.

The companies suggest that patents and patent applications that, firstly, describe "an idea with a high degree of generality" and, secondly, do not tell how the idea can be put into practice, are ineligible. The patents merely state that the design can be applied "in a computer" or "on the Internet".

If someone comes up with an "abstract idea", patent holders can sue them for infringement of their rights. Such a system "hinders rather than advances technological progress", the companies argue, because rights are "granted to those who have not come up with any significant innovations".

Eight allied companies described the signs of ineligible patents in an expert report sent to a US appeals court on Friday, 7 December. The court is being asked to dismiss lawsuits involving patents with "abstract ideas".

The opinion was prompted by litigation between two financial services companies, Alice Corp. and CLS. The subject matter of the litigation is a "data processing system for exchanging bonds between parties" that Alice Corp. patented. According to Google, Facebook and other signatory companies, the system falls into the class of "abstract ideas."

Against the background of this quite negative information, reports from also high-tech companies look much more confident, but companies that did not originate on the Internet or in social networks.

These companies create inventions based on classic principles that have guided their inventions since before the internet and social media.

Such companies do not intend to fight with anyone over the recognition or non-recognition of their patents, they simply intend to produce their innovative products and sell to patent litigants on both sides of the patent dispute.

Here are a few examples of such inventions.

IBM engineers have managed to create a set of devices that integrate electrical and optical transmission of information in a single chip manufactured using standard CMOS technology.

Among the devices that IBM has managed to develop are multiplexers that translate electronic signals into optical signals at different wavelengths, detectors that perform the inverse task, and various modulators. All of these are in the form of an integrated chip rather than individual components.

One of the devices presented is a transceiver that transmits information through an optical channel at a rate of 25 gigabits per second per channel. The device is capable of sending multiple data streams into a single optical channel by using light of different frequencies.

Similar examples can be given for other companies working in the field of patenting according to classical schemes and principles.

As a conclusion, we would like to offer our view on the formation of a technical solution, which becomes the basis of an invention.

Since the main distinguishing features take place in the components of a complex integrative product, the causal links need to be sought in the adaptation and integration of the specified innovative components in this product.

Since it can be assumed that similar components can be used in other products, in all likelihood the patent will go to the person who first used the component .

List of used literature, patent and licence materials

Annex 1
US Patent Application20190302107

A1 type code

Kauffman, Stewart, et al. 3 October 2019.

HYBRID QUANTUM-CLASSICAL COMPUTING SYSTEM AND METHOD

Annotation

Disclosed herein are systems and embodiments of systems operating between fully *quantum* coherent and fully classical states. Examples include a hybrid *quantum-classical* computing system comprising a plurality of *quantum* processors interconnected by classical means.

Annex 2
U.S. Patent Application 20190310070
A1 type code
<u>**MOWER, JACOB C. ; et al. 10 October 2019.**</u>
METHODS, SYSTEMS AND DEVICES FOR PROGRAMMABLE QUANTUM PHOTONIC PROCESSING

Annotation

A programmable photonic integrated circuit implements arbitrary linear optical transformations in a basis of spatial modes with high accuracy. Within a realistic fabrication model, we analyse programmable implementations of the CNOT gate, the CPHASE gate, an iterative phase estimation algorithm, state preparation and *quantum* random walks. We find that programmability greatly increases the robustness of the device to manufacturing imperfections and allows a wide range of experiments in both *quantum* and classical linear optics to be realised on a single device. Our results suggest that existing fabrication processes are sufficient to build such a device on a silicon photonics platform.

U.S. Patent Application20190347576

A1 type code

Von Salis, Jan R. ; et al. 14 November 2019.

MULTI-QUANTUM ENTANGLING GATE USING FREQUENCY-MODULATED TUNABLE COUPLER

Annotation

The quantum processor includes n fixed frequency *quantum* circuits having different frequencies, wherein n.gtoreq.3. The device also includes a frequency tunable coupler designed such that its frequency can be simultaneously promoted at m frequencies, wherein m.gtoreq.2, and wherein said m frequencies correspond, each, to an energy difference between a respective pair of *quantum* states spanned by the *quantum* circuits. The *quantum* circuits are, each, coupled to a tunable coupler. The method may be based on modulating the frequency of the tunable coupler simultaneously at said m frequencies. This is done, for example, in order to induce m energy transitions between coupled pairs of states spanned by the *quantum* circuits and to achieve an entangled state of the *quantum* circuits as a superposition of l states spanned by the *quantum* circuits, l.gtoreq.m.

Annex 4

U.S. Patent Application 20190347076

A1 type code

<u>**PARK, Kyung-Hwan ; et al. 14 November 2019.**</u>

DEVICE AND METHOD FOR QUANTUM RANDOM NUMBER GENERATION

Annotation

Exemplary embodiments of the present invention provide a *quantum* random number generation device according to an exemplary embodiment of the present invention, comprising: a spatially separated semiconductor detector including a plurality of cells, each cell individually absorbing a plurality of emission particles emitted by a radioactive isotope; and a signal processor that generates a random number based on an absorption event in which the plurality of emission particles are absorbed in a

Annex 5

U.S. Patent Application20190347575

A1 type code

Pedneau ; Edwin Peter Dawson ; et al. 14 November 2019.

MODELLING OF QUANTUM CIRCUITS ON *COMPUTER* USING HIERARCHICAL DATA STORAGE

Annotation

The present document describes modelling of an input *quantum* circuit, comprising a machine-readable specification of the *quantum* circuit. Aspects include partitioning the input *quantum* circuit into a group of subcircuits based on at least two groups of qubits defined for tensor slicing, where the resulting subcircuits have associated sets of qubits to be used for tensor slicing. Modelling can occur in stages, one stage per sub-circuit. The set of qubits associated with a sub-circuit may be used to partition the modelled *quantum* state tensor for the *quantum* state input circuit into *quantum* state tensor slices, and the *quantum* gates in that sub-circuit may be used to update the *quantum* state tensor slices into updated *quantum* state tensor slices. The updated *quantum* state tensor fragments are stored in secondary storage as microfragments.

Annex 6
U.S. Patent Application 20190340532
Good Cod A1
DUCORE, Andrew Mapps ; et al. 7 November 2019.
CHARACTERISATION OF QUANTUM *COMPUTER* SIMULATOR

Annotation

Various aspects of *quantum computer* simulators are described in the disclosure. In one aspect, a method of characterising a *quantum computer* simulator includes identifying simulator processes supported by the *quantum computer* simulator, generating, for each simulator process, characteristic curves for various gates or *quantum* operations, wherein the characteristic curves include information for predicting the time required to simulate each of the gates or *quantum* operations in a respective simulator process, and providing the characteristic curves to the In another aspect, a method for optimising simulations in a *quantum computer* simulator is described, wherein the simulator process is selected to simulate a circuit, *quantum* program or *quantum* algorithm based on characteristic curves that predict the time required to perform the simulation.

Annex 7
U.S. Patent Application20190332731

A1 type code

Chen, Jianxin ; et al. 31 October 2019.

METHOD AND SYSTEM FOR QUANTUM COMPUTING

Annotation

One embodiment described herein provides a system and method for modelling the behaviour of a quantum circuit that includes a plurality *of quantum* gates. At runtime, the system receives information representing the *quantum* circuit and constructs an undirected graph corresponding to the *quantum* circuit. A corresponding vertex in the undirected graph corresponds to an individual variable in the Feynman path integral used to calculate the amplitude of the *quantum* circuit, and a corresponding edge corresponds to one or more *quantum* gates. The system identifies a vertex in the undirected graph that is connected to at least two two-quantum quantum gates; simplifies the undirected graph by removing the identified vertex, thereby effectively removing the two-quantum *quantum* gates connected to the identified vertex; and evaluates the simplified undirected graph, thereby facilitating modelling of the behaviour of the *quantum* circuitry.

QUANTUM COMPUTING IMPROVES TRANSPORT

Annotation

Methods and systems for a *quantum* computing approach to solving complex, e.g., NP-complete, transport problems. One method includes (a) inputting transport data into a graph structure, wherein the transport data is associated with a transport system; (b) determining a transport metric associated with the transport system; (c) determining at least one attribute associated with the transport data, wherein the transport metric is based at least in part on the attribute; (d) using a *quantum computer* to obtain a working parameter for the attribute that improves the transport metric; and (e) applying the working parameter to operation of the transport system.

Annex 9

QUANTUM-CLOSED NANOSTRUCTURES WITH IMPROVED HOMOGENEITY AND METHODS OF THEIR PREPARATION

Annotation

The method includes: providing a substrate including a layer of crystalline material having a first surface; and exposing the first surface to an ambient environment under conditions sufficient to allow epitaxial growth of a layer of deposited material on the first surface, wherein exposing the first surface to the ambient environment includes illuminating the substrate with light having a first wavelength, causing epitaxial growth of the layer of deposited material. The first surface includes one or more discrete growth areas, wherein the epitaxial growth rate of the material having a *quantum* confined nanostructure is greater than at portions of the first surface remote from the growth areas by an amount sufficient for the deposition material to form a *quantum* confined nanostructure at each of the one or more discrete growth areas.

QUANTUM BIOS FOR RECONFIGURATION OF QUANTUM COMPUTING ARCHITECTURES

Annotation

Methods and systems for controlling an integrated optical control system for *quantum* computing using a *quantum* bios chip are described herein. A *quantum* bios chip including one or more qubit interconnect geometries and one or more error correction codes associated with the qubit interconnect geometries receives instructions associated with a *quantum* computing application. *The quantum* biochip configures one or more switching elements of an integrated optical control system coupled to the *quantum* biochip, wherein the switching elements control entanglement of one or more qubits of the quantum *computer*, and the switching elements are configured based on selected one of the one or more qubit interconnection geometries and one of the one or more error correction codes compatible with the selected one or more qubit interconnection geometries.

Annex 11

U.S. Patent Application 20190318259

A1 type code

<u>**Mohseni, Masoud ; et al. 17 October 2019.**</u>

DESIGN AND PROGRAMMING OF QUANTUM EQUIPMENT FOR RELIABLE QUANTUM ANNEALING PROCESSES

Annotation

Among other things, the device includes *quantum* units; and couplers between the *quantum* units. Each coupler is configured to couple a pair of *quantum* units according to a *quantum* Hamiltonian characterising a *quantum* using the coupler.

U.S. Patent Application20190311284

A1 type code

Mohseni, Masoud ; et al. 10 October 2019.

DESIGN AND PROGRAMMING OF QUANTUM EQUIPMENT FOR RELIABLE QUANTUM ANNEALING PROCESSES

Annotation

Among other things, the device includes *quantum* units; and couplers between the *quantum* units. Each coupler is configured to couple a pair of *quantum* units according to a *quantum* Hamiltonian characterising a *quantum* using the coupler.

Annex 13
U.S. Patent Application20190325166

A1 type code

Suresh; Vikram ; et al. 24 October 2019.

POST-QUANTUM PUBLIC KEY SIGNATURE OPERATION FOR DEVICES WITH RECONFIGURABLE CIRCUITS

Annotation

Embodiments of the invention are directed to *post-quantum* public key signature operation for devices with reconfigurable circuitry. In one embodiment, the device includes one or more processors; and a reconfigurable circuit device, wherein the reconfigurable circuit device includes a dedicated cryptographic hash hardware engine and a reconfigurable fabric including logic elements (LE), wherein the one or more processors are to configure the reconfigurable circuit device to operate the public key signature including mapping a state machine for generating and validating the public key to the reconfigurable fabric, including mapping the

The cryptographic signature generation and verification.

U.S. Patent Application 20190305206
Good Cod A1
Harris, Richard G. ; et al. 3 October 2019.

SYSTEMS, METHODS AND DEVICES FOR ACTIVE COMPENSATION OF QUANTUM PROCESSOR ELEMENTS

Annotation

Devices and methods can actively compensate for unwanted divergence in superconducting elements of *a quantum* processor. The Qbit may include a primary composite josephson junction (CJJ) structure, which may include at least a first secondary CJJ structure to provide compensation for a josephson junction asymmetry in the primary CJJ structure. The qubit may include a serial LC circuit coupled in parallel with the first CJJ structure to provide tunable capacitance. The qubit control system may include means for tuning the inductance of the qubit circuit, such as a tunable coupler inductively coupled to the qubit circuit and controlled via a programming interface, or a CJJ structure serially coupled to the qubit circuit and controlled via a programming interface.

Annex 15
U.S. Patent Application 20190327095
A1 type code
<u>**Hong; Changho ; et al. 24 October 2019.**</u>
A DEVICE AND METHOD FOR SECURE QUANTUM SIGNATURE
Annotation

An apparatus and method for a secure *quantum* signature. A method using an apparatus for a secure *quantum* signature includes preparing a *quantum* signature by sharing a first secret key and a first Bell state with a signer terminal device and sharing a second secret key and a second Bell state with a verifier terminal device; signing a message by the signer terminal device with a *quantum* signature using the first encoding value, the first secret key, and the first Bell state; and verifying by the apparatus a message using the first encoding value, the first secret key, and the first Bell state.

A1 type code

FU; Yingfang 3 October 2019.

AUTHENTICATION METHOD, DEVICE AND SYSTEM FOR QUANTUM KEY DISTRIBUTION PROCESS

Annotation

The present invention discloses an authentication method for a QKD process, and further discloses two additional authentication methods and corresponding devices, and an authentication system. The method includes the following steps: a sender selects a basis for preparing authentication information according to an algorithm in an algorithm library, and accordingly applies different wavelengths to send *quantum* states of control information and data information according to a predetermined information format; a receiver filters the received *quantum* states, uses a measurement basis corresponding to the same algorithm to measure the *quantum* state of the authentication information, and sends back authentif In addition, the sender terminates the sending process if its local authentication information does not match the reverse authentication information. In this embodiment, the authenticity of the identity of the communication participants can be confirmed in real time to effectively defend against man-in-the-middle and DDoS attacks; furthermore, the authentication information is generated based on an algorithm to prevent the waste of *quantum* keys.U.S. Patent Application20190303242

A1 type code

Kapit, Eliot 3 October 2019.

SYSTEMS AND METHODS OF PASSIVE QUANTUM ERROR CORRECTION

Annotation

Error-transparent *quantum* gates can be implemented with one or two logic qubits, each with multiple physical qubits interconnected. Error-transparent *quantum* gates implement a Hamiltonian that commutes with the single-error Hamiltonian of the logic qubits, and thus can operate successfully even in the presence of single errors. As a result, error-transparent *quantum* gates can operate with higher accuracy than their error-transparent counterparts. Each of the logic qubits can be, for example, a very small logic qubit (VSLQ) formed from a cluster of transmon or other superconducting qubits.

Buy your books fast and straightforward online - at one of world's fastest growing online book stores! Environmentally sound due to Print-on-Demand technologies.

Buy your books online at
www.morebooks.shop

Kaufen Sie Ihre Bücher schnell und unkompliziert online – auf einer der am schnellsten wachsenden Buchhandelsplattformen weltweit! Dank Print-On-Demand umwelt- und ressourcenschonend produzi ert.

Bücher schneller online kaufen
www.morebooks.shop